How to Start a Worm Bin

Your Guide to Getting Started with Worm Composting

Henry Owen

Library of Congress Cataloging-in-Publication Data

Names: Owen, Henry (Robert Henry), 1983- author.
Title: How to start a worm bin : your guide to getting started with worm composting / Henry Owen.
Identifiers: LCCN 2016036616 | ISBN 9781570673498 (pbk. : alk. paper)
Subjects: LCSH: Earthworms. | Earthworm culture. | Vermicomposting.
Classification: LCC SF597.E3 O94 2016 | DDC 639/.75--dc23

This title is printed on responsibly harvested paper stock certified by The Forest Stewardship Council, an independent auditor of responsible forestry practices. For more information, visit https://us.fsc.org.

FSC
www.fsc.org
MIX
Paper from responsible sources
FSC® C005010

Visit the author's website at
www.wormcompostinghq.com

Printed in the United States of America

ISBN: 978-1-57067-349-8

21 20 19 18 17 16   1 2 3 4 5 6 7 8 9

"Henry's book has made the art of worm composting accessible to everyone. His simple, straight forward approach, and use of everyday language makes this book a wonderful guide for the beginner worm wrangler, and a great resource for those with more experience."

—Nadine Ford, Manager,
Little Sugar Creek Community Garden

"This book is an excellent guide to vermiculture—I have never seen the process, from start to finish, explained so simply along with informative descriptions of everything from digestion to reproduction. Soon you too will be ready to pursue a new hobby making waste into high value fertilizer."

—Rich Deming, Partner,
Power Resource Group

"I found this book very informative, well written, and an easy read. It gives the do's and don'ts that are very helpful for newcomers as well as those who have already started worm composting. The science and biology involved are explained in a simple, effective format without getting overly technical. I highly recommend this book to anyone interested in a unique way to help themselves and the environment."

—Mike Gorman,
Founder, FC Organics

"This book highlights how easy and rewarding vermi-composting can be. I believe that anyone who reads this book will feel confident enough to get busy and start their very own worm farm. Vermicomposting is an easy and gentle way we all can contribute to healing the planet and this book will help guide many down this path."

—Brian Rosa, Owner/Consultant,
BE New Organic World, LLC

"Henry takes the guess work out of worm composting. He writes in a way that is easy and fun enough for a child to understand, yet engaging enough to adults turning the page. If you're tired of scrolling through the seemingly endless amateur worm composting websites or through dense university vermicompost research sites, search no more - this book is your one-stop-shop for all things worm compost."

David Valder, Co-founder
Crown Town Compost, Charlotte, NC

Your Free Gift

Thanks for purchasing my book. I'm offering a free report that is exclusive to my book and website readers.

If you are brand new to worm composting it is tough to know what equipment you really need. *Inside my Worm Composting Toolbox* is my personal list of essential tools and supplies to make worm composting easy.

You can download this free report at
http://www.wormcompostinghq.com/toolbox

Once again, thanks for your purchase and happy worm composting.

Contents

Introduction

This book is for anyone interested in learning more about worm composting or anyone ready to start their own worm compost bin. Each chapter was written to provide worm composting beginners with clear, accurate information about major worm composting topics.

You will learn how to build a worm bin, how to care for your worms, how to harvest the worm castings (poop), and how to make the best use of your valuable worm compost.

Who I Am and Why I Worm Compost

I believe I should take responsibility for my own food waste instead of sending it to a landfill. Through worm composting, I not only take responsibility for my food waste but also turn a negative (food scraps) into a positive (worm poop), which helps me close the nutrient loop and grow more food for my family. Also, worm composting is fun. I'm proud to say, "I've got worms!"

I started worm composting while working for Friendship Gardens, a nonprofit project I helped start that teaches gardening and grows food for Friendship Trays,[1] Charlotte's Meals On Wheels program. Friendship Trays serves 750 meals each weekday, and I was responsible for composting all the food waste from meal prep! Some 750 meals a day means a lot of onion peels, apple cores, and broccoli stems. We used three large flow-through worm composters and hundreds of thousands of red wiggler worms to eat through all that food waste each day. Informally and professionally, I have been teaching people to worm compost since 2008.

I have two young boys who love to play in the dirt with me and rarely try to eat the worms.

Why I Wrote This Book

I believe people are seeking ways to connect to nature that lessen the negative impact our culture has on the environment. I am passionate about teaching others about the benefits of worm composting.

This book, which includes everything a beginner needs to know to start a worm bin, allows me to be a more efficient teacher and reach a wider audience.

---

[1] http://friendshiptrays.org

Why Did You Choose This Book?

- What interests you about worm composting?
- What do you believe that makes worm composting a good fit for your life?
- Do you believe in taking responsibility for the waste we create?
- Do you love to garden?
- Do you believe in nature's ability to turn "waste" into a benefit?
- Do you believe in the power of the red wiggler worm?

Let's get started!

Chapter 1
What Is Worm Composting?

Worm composting (vermicomposting) is easy, convenient, good for the earth, good for your plants, and fun.

Worm composting allows you to responsibly dispose of your kitchen scraps while at the same time creating a high-nutrient compost that you can use on your indoor or outdoor plants.

The worms have three jobs: eat, poop, and make babies. Your job is to manage them in a way to maximize all three.

Worm Composting Is Easy

Children can easily manage a worm bin with a little supervision from adults. Once a week or so you feed the worms by burying the food scraps under the bedding. Once a year or so (you can do it more often if you need the vermicompost) you harvest the bin and restart so that you can use the good, nutrient-rich worm compost.

Worm Composting Is Fast

One of the main reasons that I started worm composting is that red wiggler worms process food scraps faster than a compost pile does and in less space. Red wiggler worms eat up to half their weight every day.

Worm Composting Is Year-round

Many people find it difficult to manage an outdoor compost pile in the winter. It is difficult to get out and turn the compost pile on cold days, and it is harder to get the pile to heat up from the inside out. You can easily do worm composting year-round by keeping the worms inside the house (I have a bin under my desk!) or in the garage.

Worm Composting Is Convenient

Worms can be kept inside or outside (in the shade). A shed or garage is a perfect spot, which also makes it convenient to add the food scraps. I know of several people who keep their worm bin in the kitchen pantry or closet to make adding food scraps even easier.

Worm Composting Is Good for the Earth

Through worm composting, you avoid throwing your family's food scraps in the trash can and sending them to the landfill to create harmful methane gas.

Worm Composting Is Good for Your Plants

Whether you grow a vegetable garden, a flower bed, a lawn, or just some indoor houseplants, your plants will benefit from the addition of your homemade, nutrient-rich worm compost (vermicompost).

Worm Composting Is FUN

Keeping worms is fun! Kids and adults love caring for worms. And it makes a great conversation starter!

Why Worm Compost?

There are two primary reasons to worm compost at home.

To responsibly dispose of your kitchen scraps. Taking responsibility for your own food scraps positively impacts the environment in several ways: it keeps food scraps out of the landfill where they create harmful methane gas, and it saves the gas and pollution needed to run garbage trucks. Imagine how few garbage pickups would be needed if no one put food scraps in the trash.

Compost is valuable! Adding compost to any plant makes it healthier. Compost is especially useful for vegetable or flower gardens, but it also makes houseplants healthier and lawns thicker and greener. Vermicompost is one of the best types of compost you can use because it is extremely nutrient-rich. Many people use vermicompost as an organic fertilizer.

What Worm Compost Does for Plants

Gardeners know that healthy plants come from healthy soil. Adding worm compost to your garden soil helps plants grow by boosting the organic matter, nutrients, and beneficial microbes in the soil. Research has shown that worm compost also suppresses diseases and insect pest attacks. Simply put: it's some good . . . worm poop.

Worm Compost Increases Organic Matter in Soil

Organic matter is simply any type of living or dead plant or animal material. Worm compost is one type of organic matter that you can add to your garden soil—it may be the best. More organic matter in your soil means enhanced soil structure, better soil drainage, and a better environment for nightcrawler earthworms to live, eat, poop, and tunnel.

Worm Compost Boosts the Nutrients Available to Plants

As worms eat and digest vegetable food scraps, the nutrients in the food are converted into forms that plants take in and use. The nutrients already exist in the apple core, but it takes a trip through a worm's digestive system to unlock those nutrients and make them usable to plants.

Worm Compost Increases the Beneficial Microorganisms in Your Soil

Microorganisms, or microbes, are tiny single-cell organisms, such as bacteria and fungi. Depending on the type, they can do both great and terrible things in your garden. By adding worm compost to your garden, you are adding beneficial microbes that can help suppress plant diseases and repel insect attacks.

Do Your Own Worm Compost Fertilizer Test

Do you want to see with your own eyes the difference worm compost can make? Do a simple test. Plant identical plants in identical pots with identical soil. Then add a bunch of worm compost to only one of them and watch what happens!

6 Stories of People Who Decided on Worm Composting

A young mom had never composted before but wanted to start. She also wanted to involve her young kids in composting. A large outdoor compost pile would have been very difficult for her kids to participate with, but they loved feeding, playing with, and showing their friends their red wiggler worms in their worm composting bin.

A couple had recently moved from their house to a condo. While in their house, they composted their food scraps in an outdoor compost pile in their backyard. Without a yard at their condo, they needed an alternative method for composting. Worm composting was a perfect fit for this couple because they could do it in a small space and use the nutrient-rich compost on their houseplants and patio plants.

After composting for years at home, a young man couldn't bear to toss the coffee grounds from the employee break room at work in the trash. He tried taking them home to compost but quickly grew frustrated with transporting wet coffee grounds (even though his car smelled great). He learned how to worm compost and started keeping a bin under his desk at work. He eventually told his boss and coworkers, and the whole office started saving the food scraps from their lunches to feed the composting red worms.

An elementary school teacher wanted to teach her students about life cycles in a creative, engaging way. Maintaining and turning a compost pile was too much work for her young kindergarten students. So the teacher and students set up a classroom worm bin that the students take care of. They love to feed the worms the leftovers from their snacks. And the teacher uses the bin as a science teaching tool.

After years of falling short, a backyard gardener wanted to finally outgrow his neighbor. He learned of the benefits of using nutrient-rich worm castings in his organic garden and decided to start a worm bin. Amending his soil with worm castings led to his best tomato crop ever, and he was hooked for life (and so was his neighbor).

A group of college students wanted to decrease their negative impact on the earth. Their research told them that to compost the large amount of food scraps from their dining hall kitchen in traditional compost piles, they would have to set up a commercial composting facility with backhoes and windrow turners to turn the piles. Or they could purchase (or make) a large-scale worm composting bin that requires less maintenance and costs much less.

If you are in a similar situation, worm composting may be a good fit for you.

Chapter 2
Fears About Worm Composting

This chapter will help calm your fears and, more important, identify solutions so you and your worms can live happily ever after.

What is holding you back from starting your own worm composting bin today? Chances are, it's one of these top five fears about vermicomposting that all beginners share.

Worm Escape

Help! Thousands of worms are crawling all over my house!

The No. 1 fear of new and experienced worm farmers is that the worms will stage a mass exodus from the worm bin. Just the thought of thousands of red wiggler worms wriggling around their house makes most people's skin crawl.

The good news is that if you take good care of your worms, this will never happen. The simple truth is that worms cannot survive outside of the habitat you have made for them, and they will not bail unless things get really, really bad inside the bin. I keep a worm bin under my desk at work and have had zero problems with it.

Following are the major reasons worms leave their bin and how to avoid them:

- ***Exploration of brand new worm composting bin:*** When you first add composting worms to a new worm bin, the worms will crawl all over the inside of the bin checking it out within the first couple days and nights. Usually they don't leave the bin during this "settling in" stage. But to make sure that they don't, simply leave a light on in the room with the worms. Worms don't like the light, so they will be encouraged to burrow down in the worm bin bedding.

- ***Too much water, not enough air:*** Another reason worms may leave their cozy worm bin is if you allow it to get too wet inside. The worm bedding needs to be moist but *not* soupy or soggy. Too much water means the excess water fills up all the space in the bedding that is usually filled with air. This lack of air and crawl space for the composting worms could cause them to look for another home outside the worm bin. If your bin is too wet, add some drainage holes to drain some of the excess moisture. Or add some dry bedding to soak up the extra moisture.

- ***Worm bin too hot:*** Composting worms will leave their bin if it gets extremely hot. Be sure your worm bin stays within the ideal temperature range by keeping it either indoors or in the shade outdoors. Also, be sure not to overfeed your worms with food scraps. A large worm composting bin can turn into a self-heating hot compost pile if you add too much food waste too quickly.

If you are still worried about your worms escaping into your house, don't give up on vermicomposting, just start an outdoor worm bin!

Smell

Rotting food scraps? Worm poop? Doesn't it stink?

As a worm farmer, you will be asked this often. The answer is: worm bins should never stink. A bad smell indicates that something has gone wrong and needs to be fixed.

Following are the major reasons a worm bin might stink and how to fix it:

- ***Too much water, anaerobic decomposition:*** There are generally two types of decomposition: aerobic (with oxygen) and anaerobic (without oxygen). We always prefer aerobic. If you have ever bagged grass clippings, tied them in a plastic bag, left them in the sun for a while, then opened the bag, you have experienced smelly anaerobic decomposition. To ensure your worm composting bin doesn't go anaerobic, be sure there isn't too much water in your bin. Water forces the air out of the bedding.

- ***Too much food waste:*** Another reason your worm bin may stink is you have added too many food scraps for your worms to eat. If the composting worms can't eat them in time, the food scraps will decompose on their own, which can cause a foul odor.

- ***Food scraps on top of bedding:*** Be sure to always bury your food scraps under the worm bin bedding to ensure they won't stink.

The Ickiness Factor

Touching one worm is fine, but holding a handful of composting worms? You have got to be kidding me!

Even though they are sold on the benefits of worm composting, many people struggle with the ickiness factor of worm farming. Some people see worms as yucky, slimy, crawly, gross.

My advice is to start slow. You don't have to hold a handful of worms on Day 1 (unless you want to). You don't even need to hold any worms in your hand until you are ready. If worms freak you out a little, use latex or rubber gloves and a small digging fork when you work with your worms.

Rats

Won't rats be attracted to the food scraps?

Nobody wants rats hanging around outside their house. Following are a few ways to ensure that rats don't see your worm composting bin as an all-night buffet.

- ***Keep the worm bin inside.*** This is the simplest solution.

- ***Bury your food scraps.*** Be sure to always bury all food scraps under the worm bin bedding. Food scraps tossed on top of the bedding can cause a slight odor, which can attract unwanted rodents.

- ***Use a sealable outdoor worm bin.*** Whether you make your own worm composting container or purchase one, be sure that your worm bin has a lid that seals it off from rodents. I have used cinder blocks to block the harvest access door in a homemade continuous flow-through worm bin.

NOTE

If you manage all other aspects of worm farming correctly, you do not need to seal off your bin from rodents. Many people have success using in-ground worm composting systems, which are essentially pits or trenches dug in the ground.

Bugs and Maggots

I don't want any other bugs in my worm bin, and I really don't want any maggots!

If you keep your worm composting bin outdoors, you will soon find that your worm bin is host to lots of other types of bugs as well as your worms. In my worm bin, I have found rolly pollies (sow bugs), ants, centipedes, millipedes, earwigs, pot worms (very small white worms), slugs, and other critters.

These other bugs won't hurt your worms and are usually a sign of a healthy worm bin. Some, like ants and black soldier fly larvae, usually won't hurt the worms but should be dealt with for your comfort.

Knock on wood—I have never had fire ants in any of my worm bins. I have had lots of sugar ants, especially in my larger outdoor worm composting systems. An ant infestation is usually a sign that your worm bin bedding is too dry. Add some water with a spray bottle or watering can. If that doesn't work, try sprinkling ground cinnamon on top wherever the ants are. Ants hate cinnamon. Don't use a poison to get rid of ants because that will hurt your worms.

People often freak out when they first find black soldier fly larvae in their worm bin because they look like maggots. They are similar to common housefly maggots—they are the larvae stage of a type of fly. However, they eat any type of food scraps, not only meat. Many people intentionally use black soldier fly larvae to compost their food scraps. I have even set up black soldier fly composters on top of my chicken coop that the adult larvae crawl out of (when they are ready to turn into a fly) and drop right down into the chicken coop. My chickens love eating them and have learned to wait under the composters. A free protein source! Black soldier fliers are found naturally in most parts of the world and will most likely find your outdoor worm bin in the summer. They will not hurt your composting worms and can easily coexist in the same bin.

If you are worried about other critters finding your worm bin, try keeping thc worm bin inside from the beginning. Rolly pollies and a few other harmless bugs will still find your bin, but you will be sure to keep out ants and black soldier fly larvae.

Chapter 3
The Worm

The red wiggler worm is the rock star of worm bins.

There are thousands of different varieties of earthworms. But if you are looking to start a worm composting bin, red wiggler worms are the clear choice for most worm farmers. In this chapter, you will learn a little about this favorite earthworm and what traits make it the best composting worm.

What to Call Them

Red wigglers (my favorite name for them because, well . . . they're red and they wiggle) are called by many different names. Their scientific name is *Eisenia fetida,* but they are also referred to as red worms, tiger worms, manure worms (they're often found in manure piles), and brandling worms.

Where Red Wigglers Live

In nature, red wigglers live in the topmost layer of soil and decaying organic matter, such as leaves and wood. They eat the decaying organic matter and microorganisms that are also feeding on the decaying organic matter. Unlike nightcrawler earthworms, which can burrow many feet down into the soil, red wigglers do not create tunnels and stay in the top 3 to 6 inches (7.6 cm to 15.2 cm).

Red wigglers were originally native to Europe but now can be found all over the world. It's possible in some areas to find red wigglers in your backyard and add them to your worm composting bin. However, searching for them is very time-consuming (I've done it!) and usually results in a small number of worms. I recommend starting with 1,000 worms to get your worm herd off to a good start.

Why Red Wigglers Are the Best Composting Worms

- ***They reproduce quickly.*** How quickly? It takes a baby worm only nine weeks to reach maturity and start reproducing. Worms reproduce by creating small tan-colored cocoons. Each cocoon holds two or three tiny baby red wigglers. So, given ideal conditions, enough food, and enough space in the worm bin, you can expect your worm herd to double every three or four months.

- ***They naturally live very close to each other.*** This is important because you want to keep a bunch of them confined to a small space, the worm bin.

- ***They tolerate a fairly wide range of temperatures.*** Red wigglers are most efficient (eating, pooping, making babies) at temperatures that we humans prefer, roughly 60°F to 80°F (15.5 C to 26.6 C).

- ***They tolerate being pawed through by worm farmers.*** This is important because you will want to check in on your worms occasionally. You may also have a young child worm farmer in your house who likes to check on them, oh, every hour or so.

- ***They eat a lot!*** Red wigglers are voracious eaters. Depending on the conditions in the worm bin, they can eat between ¼ and ½ of their weight every day. So, if you have 1 pound (453.6 g) of worms (roughly 1,000 worms), you can expect them to eat ¼ pound (113.4 g) to ½ pound (226.8 g) to scraps each day—under ideal conditions.

How Worms Eat

As worm farmers, we know that our composting worms have three jobs: eat, poop, and make babies. We know that eating is obviously important, but how do composting worms actually eat?

Digestive Anatomy of a Composting Worm

Composting worms do not have teeth. Instead, they grind their food in small gizzards. The gizzard (birds have one too) is a small sack early in the digestive tract (which for a worm runs the entire length of its body) that contains very small bits of grit or sand. The food passes through the gizzard and gets ground up by the grit. For this reason, I recommend adding a small amount of grit to your worm bin with you first start. Good sources of grit are dirt from your yard, rock dust, or crushed oyster shell.

Without teeth, worms cannot take a bite out of food. They need to wait until the food begins to rot or break down so that it is soft and wet enough for them to suck off with their very small mouths. Remember, your composting worms eat food scraps, worm bin bedding, and tiny bacteria that are working to break down the food scraps. The more surface area (smaller pieces) the food scraps have, the faster the worms and bacteria can eat it.

How to Make It Easier for Your Composting Worms to Eat

- *Make more surface area* in your worm food by chopping it up.

- *Make some slurry.* Run your food scraps through a blender before adding them to your bin. This increases surface area and makes a nice, soft, mushy slurry that is easy for your worms and bacteria to eat.

- *Freeze your food scraps* before adding them to your worm bin. Freezing food scraps breaks down the walls of individual cells, helping them break down faster once they thaw out.

- *Microwave your food scraps.* This not only softens food so that it starts to break down faster but is also a great way to make sure that you kill any fruit fly larvae so that you don't accidentally add them to your bin—fruit flies are the worst!

Worm Reproduction

Making new worm babies is one of your worms' three key jobs.

Reproductive Anatomy of an Earthworm

Earthworms are hermaphroditic: they are neither male nor female. Instead, each worm has a complete set of male and female reproductive organs.

This does not mean that worms can reproduce by themselves. They need to find another sexually mature worm to mate with. Any mature worm can reproduce with any other mature worm in the bin. This may sound strange to us as humans, but it makes increasing your population very easy because each worm has so many more potential partners.

Sexually mature worms can be identified by their clitellum, the thick band located near the worm's head. If you have ever played with large nightcrawler earthworms in your backyard, you know what I'm talking about. The clitellum is the earthworm's reproductive organ.

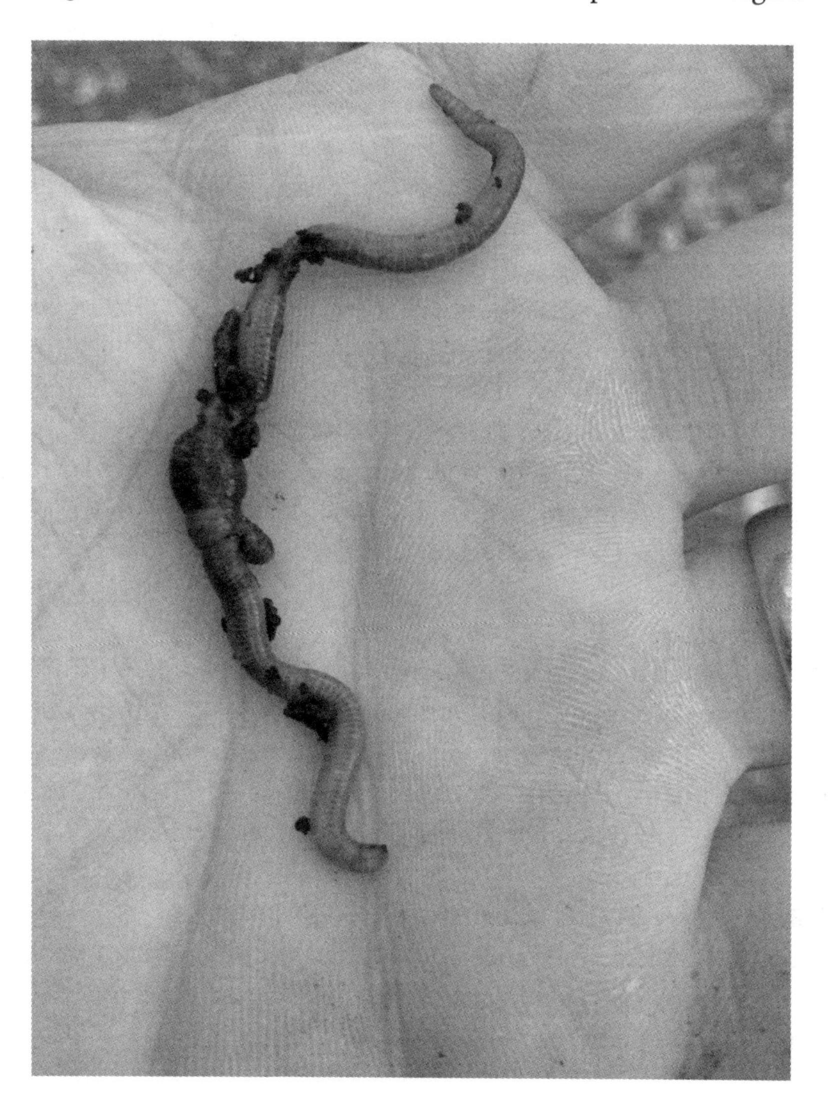

How Worms Make Babies

The reproductive act, which lasts for around three hours, begins with two mature worms giving each other a hug. They line up their clitellum and then hold on to each other's bodies with tiny hairs called setae. During this hug, the worms swap reproductive seminal fluids.

Next, the worms secrete mucus rings around their bodies. The visual effect is that it makes the worms look like they are tied up with very fine fishing string. These mucus rings are the beginnings of the shell of the worm cocoon.

When the worms separate, the mucus ring slides off each worm, collecting fertilized reproductive seminal fluids as it moves along the worm's body. When the mucus ring gets to the end of the worm, the ends of the mucus ring seal themselves, creating the cocoon that contains all the necessary reproductive material. The cocoon then separates from the worms to develop.

Red Wiggler Earthworm Cocoon

The red wiggler worm cocoon is small and lemon-shaped. The hatchlings inside the cocoon grow for a little over three months before they hatch. Each cocoon usually hatches two to four baby red wiggler worms.

Amazingly, worm cocoons can hibernate. If conditions in nature or in your worm composting bin are not stable enough for baby worms to survive (extreme temperatures, low moisture), hatching is delayed until conditions improve. Once the conditions do improve, the reproductive cycle is completed, and the baby worms emerge.

Baby Red Wiggler Composting Worms

When they emerge from their cocoons, red worm hatchlings are very tiny and white. In roughly one day, the white color begins to transition to the red that they will have as adults.

Don't confuse hours-old white baby red wiggler worms with pot worms, which are also small and white. Pot worms are another variety of worm that lives happily in a worm bin or compost pile. Though pot worms aren't by themselves a bad thing (they also help with decomposition), they thrive in different conditions than red wigglers do. So, if you have a pot worm infestation, your worm bin or compost pile is not set up well for your red wigglers to thrive.

In nine weeks, your baby red wiggler worms will grow to maturity and be ready to start the worm reproduction cycle all over again.

Where to Buy Red Wiggler Composting Worms

You can buy composting worms online or from a local source. Good worm producers[2] know how to package and ship worms so that they are guaranteed to arrive alive at your door. Alternatively, you can try to find a worm composting friend willing to share some worms with you to get you started.

Consider the following tips before you purchase composting worms from a seller:

- ***Don't buy from a bait shop.*** Many bait shops and hardware stores sell red worms in small quantities as fish bait. Do not purchase these worms to stock your worm composting bin. Because they are sold as bait, these worms are often much more expensive than purchasing composting worms by the pound. Worms raised for fishing are usually fed a specific diet to fatten them up. They are best suited for the end of a fishing line, not for your worm bin.

- ***Make sure the seller guarantees live delivery.*** If not, do not purchase. Any good composting worm seller will guarantee that his worms will be alive when they arrive at your door. Reputable worm sellers do not ship during weeks of extreme

---

[2] http://www.wormcompostinghq.com/buyworms

hot or extreme cold temperatures. A box of dead worms never helped anyone.

- ***You get what you pay for.*** As with most products, the cheapest option is usually a bad product. The cheapest-cost worm sellers may skimp on the amount of worms, package in poor quality materials, and offer no customer support. Going with the cheapest option is always a gamble.

Navigate to http://www.wormcompostinghq.com/buyworms to see my experienced worm seller recommendation.

Chapter 4
Setting Up Your Worm Farm

Now it's time to measure your family's food waste, calculate how many composting worms you need, and determine what size worm bin you need for composting.

Your Family's Food Waste

Before you can calculate how many worms you need, you need to track the average amount of veggie food waste that you or your family creates.

Following are simple instructions for how to track food waste and conduct a family food waste audit:

- Collect a day's worth of vegetable food waste that could have been fed to your worm farm.
- Be sure to collect only the types of food waste that worms can eat.
- At the end of the collection day, place all veggie scraps in a plastic bag and weigh it. A small kitchen scale or fish scale works well.
- Repeat for one week. Then average the results together.
- You will now have your daily average food waste for your household.

Number of Worms You'll Need

Amazingly, red wiggler composting worms eat roughly half their weight every day. So, if your daily average food waste is 2 pounds

(907.2 g), you will need roughly 4 pounds (1 kg 814.37 g) of composting worms to eat that amount each day. In this scenario, 4 pounds of worms is your optimal worm herd.

FORMULA FOR OPTIMAL WORM HERD SIZE

average daily food waste (by weight) X 2 = amount of composting worms (by weight)

Example: 2 pounds (907.2 g) food scraps, 4 pounds (1 kg 814.37 g) of worms

Size of Worm Bin You'll Need for Your Optimal Worm Herd

Assuming that temperature and moisture conditions are ideal and there is plenty for a worm population to eat, your worm population will expand to the size of its habitat (worm bin and bedding). Composting worms feed in the top inch of worm bedding. So, the next step is to calculate how many square feet you need on the surface of your worm bin.

One pound (453.59 g) of red wigglers for 1 square foot (0.09 square meters) of surface area in a worm farm is a good, healthy ratio (1:1). So, if your optimal worm herd is 4 pounds, you will need a worm farm with at least 4 square feet of surface area. Many plastic storage bins fit this size and make excellent worm farms.

FORMULA FOR SURFACE AREA OF WORM BIN

amount of worms (in pounds) = surface area (in square feet);

453.59 g of worms = 0.09 square meters of surface area

Example: 4 pounds of worms, 4 square feet of surface area.

Since all of our calculations are based on weight, you don't actually need to know how many worms are in a pound ((453.59 g). However, just because it's fun to know roughly how many worms are in your worm herd, most worm composters use the estimate of 1,000 worms in 1 pound. Therefore, if you have 4 pounds (1 kg 814.37 g) of worms in your vermicompost bin, you are feeding around 4,000 worms!

In our example so far, you need 4 pounds of composting worms to eat all the food scraps that your family produces. To get up to 4 pounds of red worms, you can do one of two things. You can buy 4 pounds of composting worms. This is the simplest but also the most expensive way to reach your optimum worm herd. Another approach is to start your worm farm with ½ pound or 1 pound of worms and let them reproduce. This is a slower approach but will save you money.

Chapter 5
The Bin

Learn about the all-important home for your worm herd.

Your composting worms need some sort of container to call home. This will be your worm bin. Almost any container can be turned into a worm bin. Or you can build your own to meet your needs and those of your worms. There are also many different models of worm bins available for purchase.

How to Build a Homemade Worm Composting Container

This section will give you recommendations on what type of container or material to use and how to build a homemade worm composting container.

Factors to Consider When Building Your Homemade Worm Composter

- *Air flow:* Worms breathe through their skin and need air to live. You need to allow your composting worms access to fresh air. You can do this by simply drilling holes in the sides of the worm composting container or bin, or by installing some sort of vent (in a larger vermicomposting system). You could also leave your worm bin open on top, provided you are planning to keep it inside a house, shed, or garage.

- ***Inside or outside?*** The answer to this question will determine whether you need to think about insulating your bin in some way to make sure the worms stay within their ideal temperature range. Too much heat and freezing temperatures kill red worms. If you choose an outside bin, build foam or Styrofoam insulation into the walls of the worm bin. Or go with the low-tech method of simply piling hay bales around the worm composting container in the winter.

- ***Size:*** The amount of food scraps you want to compost should determine the size of your worm composting container. The size of the composting container (more specifically, the volume of the bedding in the worm bin) will determine the size of your worm population. The size of your red wiggler worm population will determine how much and how often you feed them. Use the formulas I mentioned above.

- ***Harvest method:*** Some worm composting containers make it easier to harvest the worm castings than others. If you choose to use some sort of storage tote, bin, or box, it will be a bit of a challenge to separate your worms from the vermicompost that you want to add to your garden. In contrast, a flow-through worm digester is more difficult to set up but makes harvesting much simpler. In a flow-through digester, the worm bedding rests on a grate raised off the ground. The worms eat upward and leave their nutrient-rich castings behind on the bottom level, which then falls through the grate—hopefully worm free—where you can easily scoop it up and use it however you want. You can easily turn a trash can into flow-through worm digester. Simply cut a hole in one of the sides near the bottom and install a grate above the hole to support the worm bedding. The very simplest way to harvest is not to harvest at all. Simply dig a hole or trench right where you want the nutrient-rich castings to eventually be, like in your vegetable garden. Then fill the hole with bedding, worms, and food scraps. Use a board as a cover, and manage the same way you would a bin or other worm composting container. Instead of needing to harvest the castings to add to your garden, the nutrient-rich worm poop will seep into the ground around your in-ground worm composter. I rotate the in-ground worm composter spot each season in my garden so that eventually the whole garden will be enriched.

Container for Your Homemade Worm Composter

- ***Anything:*** As long as you follow correct vermicomposting principles, almost any container can be used as a homemade worm composting container. I've had success using storage totes, wooden boxes, plastic containers, an old chest freezer, a cracked ice chest, and several rolling trash cans.

- ***Storage tub:*** Rubbermaid-style storage totes/tubs are most commonly used by beginners. They are inexpensive and simple to get started. You will have to drill holes in the sides for air flow.

- ***Small plastic tub:*** Some families keep a small plastic worm composting tub in their kitchen or under their sink for easy access to food scraps. Using a clear tub is fun because you can see the worms moving around as they work.

- ***Wood:*** Wood is cheap and easy to find. Because the bedding needs to be wet, the wood will rot in a handful of years. Be sure not to use painted wood on the inside of the bin because this could affect the worms.

- ***Old freezer or ice chest:*** This works particularly well because it already has built-in insulation.

- ***Trash can:*** A trash can makes an excellent flow-through worm digester or large simple worm bin. If you opt for the rolling trash can, you have the added benefit of its being easy to move.

- ***No bin, dirt walls:*** As I mentioned above, you can worm compost in the ground right where you want the benefits of the nutrient-rich worm castings. I use this technique in my vegetable garden, but you could easily also do this in your flower bed.

If you follow good vermicomposting principles, you can convert almost any container into a homemade worm composter. Chances are you have some sort of old storage bin or ice chest right now that you could use. Or, using the above information, you could design and build your own worm composting bin.

How to Make a Simple Starter Worm Bin

Most new worm farmers prefer to start with a simple storage tub-style worm bin. They are cheap, easy to make, and easy to manage.

To make your homemade worm bin, you will first need to gather your tools and materials:

- One 14-gallon (53 L) to 20-gallon (75.7 L) storage tub with a lid
- A drill and drill bit (any size close to ½ inch (1.27 cm))
- *Lots* of shredded newspaper (way more than you think!)
- A bucket of water
- ½ pound (226.8 g) to 1 pound (453.6 g) of red wigglers (*Eisenia fetida*)

Step-by-Step Instructions for Building a Worm Bin

1. Use the drill and drill bit to drill holes in the sides of the storage tub. The storage tub will be the bin that the worms live in. Remember, worms need to breathe. Six to ten holes on each side are enough. The worms will not crawl out the holes because they know they will die outside of their moist worm bedding.

2. Shred or rip up the newspaper. Your red wiggler worms need bedding to live in. Any carbon source will work, but use shredded newsprint (no glossy paper). Rip the newsprint into 1- to 2-inch (2.5 cm to 5 cm) strips.

3. Red wigglers need a very moist environment to live. Build your worm bed with shredded, soaked newspaper. Put all the ripped newsprint strips into the bucket of water, and submerge them.

4. Wait five minutes for the newsprint to soak up the water.

5. Fill the storage tub with the shredded, soaked newspaper. Pick up handfuls of the newsprint, and let the water run out of it for 10 seconds or so before adding it to the bin. Red wriggler worms like a moist environment, like a wet sponge. They don't like swimming. There should never be standing water in the bottom of your homemade worm bin.

6. Fill the worm bed at least half full with the shredded, soaked newspaper.

7. Add a handful of dirt or homemade compost. The dirt or compost will give them tiny bits of grit and sand that they can eat and use in their gizzards to grind up the food scraps that you will feed them.

8. Add your worms. I recommend starting a homemade bin of this size with ½ pound to 1 pound of worms. The worms will multiply. Remember, one of their three jobs is to make babies. So, the amount you start with really depends on how productive you want your homemade worm bin to be right away, and how much you are willing to spend.

9. Put the lid on the bin and leave it alone for two days in a lighted room or closet. Earthworms like to explore their new home. In the first two days, they will crawl all over the inside of your worm bin. Worms don't like light, so leaving a light on in the room of closet encourages them to stay in the bin. After two days, if you have taken good care of them, they will never leave the bin.

10. Feed your worms. Pull some of the bedding back to one side, and add some food scraps. Then cover the food scraps with bedding. Remember, worms are vegetarians; don't feed them

any meat. Be careful not to overfeed them. Start with a small amount and monitor how fast they eat it. When it is gone, or mostly gone, add some more.

11. Feed more. Feed your worms once or twice a week, depending on how much you feed them each time.

12. Harvest the vermicompost (worm poop). It will take around six months before you are able to harvest the first time. After that, you can harvest more frequently. Start by simply digging down to the bottom of the bin and grabbing a handful of the vermicompost. You can either add that to your garden or pick out the worms and then add it to your garden.

Buying a Worm Bin: Worm Bin Reviews

If you would rather purchase a worm bin, there are many available to choose from. Which commercially available worm bin is right for you? I have reviewed a few of the worm bins on sale now.[3] The information below is based on my personal experience with each worm bin.

Can-O-Worms

This is my favorite worm bin available for purchase. Like many worm bins, the Can-O-Worms uses a tray system that allows for flow-through worm composting. Each tray has many holes in the bottom to allow the composting worms to move up and down the system while keeping the vermicompost in the tray. Once the bottom tray in the stack has been fully processed by the worms, you simply remove that tray, use the worm compost, fill the empty tray with fresh bedding, and add it to the top of the system.

Because composting worms live in the top few inches of bedding, they are constantly moving upward through the holes in the trays to find fresh bedding and food, leaving their valuable worm castings

[3] http://www.wormcompostinghq.com/worm-bin-reviews

behind in the lower trays for you to easily collect. This system also includes a valve at the bottom so that you can remove excess worm bin leachate (liquid created in the worm composting process). If you are going to purchase a worm bin, this is the one I recommend.

What I Like About the Can-O-Worms

- ***Large trays:*** Each tray has a surface area of 314.16 square inches (2033.3 square cm). Because red wiggler composting worms like to live in the top few inches of worm bin bedding, surface area of your worm bin is more important than depth. If you want a larger worm herd that can process food scraps into worm castings faster, you need to optimize the surface area in your bin. The Can-O-Worms comes with three large circular flow-through trays.

- ***The round shape:*** I think the round shape looks better than a square. And the circular trays help maximize the surface area of each tray.

- ***The lid:*** unlike some of its competitors, the Can-O-Worms includes a lid with a screened air vent that fits snugly on top of the system.

- ***The height of the system:*** You can easily fit a collection container under the valve to collect the worm bin leachate. The drain valve is 8 inches (20.32 cm) above the floor.

What I Don't Like About the Can-O-Worms

- ***Fragile legs:*** The legs aren't as sturdy as I would like. I am nitpicking here, but I would rather the legs of the system be a little sturdier. I have not broken one of the legs, but when the system is full (all trays filled and active) I have worried a bit that one would break, especially when trying to move the full worm bin.

The Worm Factory

This was the first worm bin that I ever purchased. I was happy with it until I found the Can-O-Worms. The Worm Factory is similar to the Can-O-Worms with a few key differences. The Worm Factory relies on a tray system that allows for a flow-through design and easy worm casting harvesting. It includes a tap at the bottom for easy worm leachate removal.

What I Like About the Worm Factory

- ***Its sturdy plastic base:*** The base feels very sturdy, even when the system is full. Overall, the system and trays are made of very sturdy, thick plastic.

- ***The opportunity to purchase extra trays:*** I purchased five total (as many as I could) when first purchasing this unit. It's nice to have an extra tray or two to work with for harvesting and other worm composting activities. The system can really handle only three trays at a time.

What I Don't Like About the Worm Factory

- ***Limited surface area:*** Each tray on the worm tower has a surface area of 210.25 square inches (1356.4 square cm). As we know, bedding surface area is more important than bedding depth.

- ***Too many trays:*** I mentioned above that I like having an extra tray or two to work with for storing worm compost or using one as a filter. But I learned quickly that using all five at once on the Worm Factory does not work. It gives the worms too much depth of bedding and food scraps. They ended up moving upward too soon and not fully processing the lower trays of food and bedding.

- *Location of leachate tap:* The worm leachate tap is too close to the ground. The drain valve is only 5.5 inches (13.97 cm) from the ground, making it more difficult to fit some collection containers underneath.

- *The lid:* The lid on the Worm Factory is a single flat piece of plastic that nests inside the topmost tray. Even though this lid does its job of blocking light, it does not seal as well as the Can-O-Worms lid does or look as nice.

The Worm Wigwam

The Worm Wigwam is a much larger worm bin. It's a good fit for a small business or school. I purchased the Worm Wigwam for work to help compost food scraps from a Meals On Wheels program. The Worm Wigwam has a flow-through design but does not use trays. Instead, fresh bedding and food scraps are continually added to the top of the system, and the worm compost is harvested from the bottom.

The worm bedding sits on a grate with a blade attached to it. A crank on the side of the Worm Wigwam moves the blade across the bottom of the worm bedding scraping off the bottom layer of worm compost and allowing it to fall down through the grate into the collection pan. Because the composting worms move upward to eat, the worm compost that you harvest from the bottom should be free of worms.

What I Like About the Worm Wigwam

- *Harvest bar:* This makes it very easy to collect valuable worm compost whenever you need it.

- *Capacity:* The Worm Wigwam is one of the few medium to large-scale worm bins available on the market.

- ***Insulation:*** There is a thin layer of insulation sandwiched between the two plastic walls of the bin. This insulation is very helpful. My only wish is that there was more of it. I keep my wigwam outdoors under shade. More insulation would help keep it warm in the winter and cool in the summer.

What I Don't Like About the Worm Wigwam

- ***Price:*** The Worm Wigwam is expensive.

- ***Assembly:*** Because it's a large system, more assembly is required of the user.

- ***Air flow:*** I found that the system could use better air flow. I ended up drilling small holes in the lid to allow for greater air flow.

- ***Productivity claims in the marketing:*** The Worm Wigwam marketing claims it can process 10–15 pounds (4 kg 535.92 g to 6 kg 803.89 g) of food scraps and bedding per day, and that it can produce 45–60 pounds (20 kg 411.66 g to 27 kg 215.54 g) of finished worm compost every week. I have not found this to be true. I practice a low-maintenance worm farming approach. Perhaps if I worked harder to make the worms' lives easier, the Worm Wigwam would approach its marketing claim numbers.

Remember, you can always build your own worm bin. But if you want a quick-start, relatively attractive, low-maintenance worm bin, give one of these a try. For the home worm composter, I recommend the Can-O-Worms.

Chapter 6
The Bedding

This chapter gives you an overview of which worm bin bedding material works best and where to get it.

Composting worms need moist bedding to survive. Almost any carbon source can be used as worm bin bedding, but some worm bedding material works better than others.

What to Use as Worm Bin Bedding

- ***Shredded paper*** makes an excellent worm bin bedding. I prefer to use shredded newspaper, but any type of shredded paper will work. Avoid the glossy section of the newspaper or the glossy junk mail. Shredded (ripped up) newspaper has enough bulk to create room for airflow and for the worms to wiggle around in. It does a great job of absorbing and holding in moisture, which composting worms need to survive. Another benefit is that you can easily find a free source of paper. If you don't get the newspaper at your house, perhaps your neighbor does. You may also have a home paper shredder that you use to dispose of sensitive information. This makes great worm bin bedding. In fact, I take great pleasure in feeding my red worms my paid bills and unwanted credit card offers!

- ***Shredded corrugated cardboard*** is another good source of worm compost bedding. Cardboard is thicker than paper, so it takes longer for the worms to process it (eat it and poop it out). Corrugated cardboard does a great job of leaving room for air and worm movement because of the space in each piece of cardboard. You can probably find a free source of cardboard. One large box will fill a worm bin. One drawback to using corrugated cardboard as your primary source of worm bin bedding is that it is difficult to shred or tear. You will make your hands sore tearing it up.

- ***Dead leaves*** can also be used as worm bin bedding. They are free and plentiful, especially in the fall. One benefit is that they don't need to be shredded or torn up. Leaves usually don't hold as much moisture as cardboard or paper does, so you may need to add water to keep the composting worms at the optimal moisture level. One drawback to using leaves as bedding in your worm bin is you will surely be adding some other little creatures and bugs that were living in the leaves. This is not a problem for the worms, but it may be for you, especially if you keep your worm bin inside. Avoid using magnolia leaves. They are too large and waxy to be used as worm bin bedding.

- ***Hay or straw*** works well as worm bedding material; however, unless you live on a farm, you will need to purchase the hay or straw. Another drawback is that hay and straw do not hold moisture as well as paper and cardboard do (after they have been soaked). Hay and straw do provide great space for air and worm movement because of their structure.

- ***Coconut coir*** is another option for worm bin bedding. Coir is a natural material made from the husk of the coconut shell. If you want to use it, you will have to buy it. It holds moisture very well and is fluffy and nice to work with. You may have received a brick of it with your initial purchase of worms, depending on where you ordered from. Coir works extremely well as worm bin bedding. I don't use coir simply because I don't want to have to purchase anything for my worm composting bin.

- ***Half-finished piled compost*** can also be used as worm farm bedding. If you have an outdoor hot compost pile already, this is another free option. You want to use material from your hot compost pile that is about halfway finished. Once it is totally finished, it is too much like dirt. Remember, red wiggler composting worms live above the soil and cannot tunnel through dirt like nightcrawlers can. To use it as worm bin bedding, pull some half-finished compost out of your pile, add some water, and add it to your worm composting bin. It will be mostly leaves, so it will have some extra critters in it. I choose not to use hot-pile compost in my worm bin because I would rather wait for the pile to finish composting in place and then use it directly in my vegetable garden.

Chapter 7
The Food

Worms eat the tiny, invisible bacteria that grow and feed on the food scraps that you add to your bin.

In nature, the red wiggler composting worm lives above the soil in wet piles of leaves or rotting wood. In your worm bin, you want to recreate this environment and get the worms to work for you.

How Composting Worms Eat

Remember, worms do not have teeth. Instead of chewing, they grind up their food in their gizzards. The gizzard (birds have one too) is a small sack early in the digestive tract that contains very small bits of grit or sand. The food passes through the gizzard and gets ground up by the grit.

Because they don't have teeth, worms cannot take a bite out of their food. They need to wait until the food begins to rot so that it is soft and wet enough for them to suck off with their tiny mouths.

Composting Worms' Favorite Foods

I have found that composting worms prefer some vegetable scraps over others. Following is a list of foods are some of my worms' favorites:

- *Melon rinds:* Cantaloupe, honeydew, watermelon. They love sweet foods with soft flesh.

- *Noncitrus fruit:* Berries, apples, pears. Be careful with citrus because it will burn their sensitive skin.

- *Squashes:* The soft flesh is easy for them to eat.

- *Slurry:* If you really want your worms to love you, throw your food scraps in the blender before adding it to your worm bin. This is certainly not necessary (I rarely do it), but the increased surface area and soft mushy slurry make it much easier for the composting worms and beneficial bacteria to eat.

Foods That Worms Do Not Like as Much

Red wiggler worms will still eat these foods, but in large quantities they could harm your composting worms.

- ***Citrus fruits:*** oranges, limes, lemons. Large quantities of citrus can burn the worms' skin.

- ***Onions and garlic:*** These can also burn their skin in large quantities.

- ***Bread:*** Composting worms will eat bread, but be careful. Too much bread in the worm bin begins to mold before the worms can get to it.

What Not to Feed Your Worms

Some things you should never put in a compost bin. Even though some of the things on this list seem natural and may even be organic, your composting worms will not be able to handle them.

- Meat (Remember, composting worms are vegetarian.)
- Dairy products
- Oil
- Cooked food (Cooked food often has seasonings, especially salt, which can harm your worms. To responsibly dispose of your cooked food scraps, and get some great eggs, get some backyard chickens.)
- Dog or cat poop (Dog and cat poop is toxic. Flea-killing meds can be especially toxic.)
- Anything greasy
- Bones
- Poisonous plants—duh
- Diseased plants
- Charcoal ash from a BBQ grill

Some Unusual Foods to Feed Your Worms

Composting worms are great at processing vegetable food scraps into nutrient-rich worm castings. Check the list below to learn nine unusual things that you currently throw away that you can start feeding to your composting worms.

- ***Dryer lint:*** It is made up of mostly plant-based fibers from your clothes.
- ***Egg shells:*** They take a very long time to break down and greatly increase the calcium content of your worm compost.
- ***Paper towels:*** Add them to your worm compost bin only if they have been used to clean up drink or food spills. Do not put paper towels that have chemicals on them in your worm bin.
- ***Pet hair:*** Be careful with this one. In small quantities, I have found that it works well. In large quantities, pet hair can easily clump together making it harder for the composting worms to break it down.
- ***Tea bags and coffee filters:*** Go ahead and throw them in as well. They are just paper (carbon).
- ***Floor sweepings:*** The contents of your dustpan is mostly dirt, pet hair, dust bunnies. They can all be added to a worm bin.
- ***Cooked pasta and rice:*** If you don't have any backyard chickens to eat up your leftover cooked pasta or rice, you can feed it to your composting worms. Be sure it doesn't have sauce or oils on it.
- ***Beard clippings:*** Similar to pet hair, beard clippings are simply dead skin cells. They are organic (used to be alive). Throw them in the bin.

- ***Sawdust from untreated wood:*** Sawdust from untreated wood is simply little bits of carbon. It will act as bedding in a worm bin. If you have a lot of sawdust to add, it's best to mix it in well with the other beddings so that it doesn't stick together.

Tips on Feeding Your Composting Worms

- ***Wait until your worms have finished their food before you feed them again.*** This is easily done by remembering where you buried the food scraps and checking that spot before feeding again.

- ***Feed an indoor worm bin more frequently with smaller amounts of food waste.*** If you keep your worm bin indoors, you'll want to manage it a bit more carefully to ensure that you never get fruit flies or foul odors. Check your indoor bin weekly, and usually feed weekly.

- ***Feed an outdoor bin once every two or three weeks.*** If you keep your worm bin outdoors, you can feed the worms a little more at each feeding and go a little longer between feedings.

- ***Be careful not to overfeed your worms.*** If you add too much food for your worms too quickly, they will not be able to eat it before it rots. Rotting food attracts fruit flies and causes a bad odor. Another way to avoid fruit flies and odor is to be sure to always bury your food scraps under the bedding.

- ***You do NOT need a wormsitter.*** If you go out of town (even for a couple weeks) your worms will be fine. Be sure to feed them before you leave. If they are outdoors, feed them a little more than usual. Remember, worms will eat their bedding as well.

- ***Be careful with garden waste/plant refuse.*** Garden waste that is rotting veggies you can simply add to the bin as you would any other worm food. You can feed plant refuse and trimmings to your composting worms, but you will need to do some work to prepare them beforehand. Often plant refuse (end of season tomato plants, small branches, stems, stalks) is very large and woody. For your composting worms to eat it, you will need to chop it up into smaller pieces. Also, consider soaking the chopped-up pieces in water overnight. Your composting worms will treat your garden waste/plant refuse like worm bin bedding. They will slowly eat it, but it will take a long time. Waste in your worm bin needs to start breaking down (rotting) before your composting worms can start to process it. (Another option for your large woody plant refuse/garden waste is to start a large outdoor hot compost pile. There are some instructions on how to start a hot compost pile at: http://infobarrel.com/how_to_build_a_hot_compost_pile.)

Where to Get More Food Scraps for Your Composting Worms

If you are trying to maximize your worm compost output or grow your worm herd, you will need to find a source of worm food beyond the food scraps that your family produces.

Here is a list of free or inexpensive sources of large quantities of worm food:

- *Coffee shops:* The added bonus here is that your worm bin will smell like delicious coffee! Most Starbucks coffee shops already bag up their spent coffee grounds for composters and gardeners. All you have to do is ask for it. If you have a favorite local shop, ask them if they would be willing to share. I ended up leaving a bucket with my name and cell phone number on it at the coffee shop near my house. When the bucket is full, they call me to come get it.

- ***Pulp from a juice bar:*** Juice pulp has the added benefit of being shredded, which makes it easy for the worms to eat.

- ***Horse stables***: Most horse stables would be happy to part with some manure, especially if you offer to help clean it out of the stales. A word of caution here—check with the owner to see what medication the horses are on. Some horse meds don't work well with worms or with the vegetables you may want to grow with your worm compost.

- ***Restaurants:*** Restaurants can be a good source of worm food. But make sure you are clear with the owners that you want only precooked food scraps. Food waste (leftovers) may contain oils and salts that are harmful to composting worms. Also, be careful about quantity. If you get a restaurant to commit to sorting its precooked food scraps (onion skins, carrot tops, ends of zucchini, etc.), it may want to give you all of them, which you may not be ready to handle. A small local restaurant may be willing to work with you for smaller quantities. You may want to start by asking the restaurant to do a food waste audit to determine how much food waste it produces. Then you can decide if you have enough worms to handle that amount.

- ***Your neighbors:*** Your family may not produce enough food scrap worm food for your growing worm herd or microworm business. But two or three families might. Ask your neighbors if they would be willing to collect their food scraps and share. You could even offer to share some finished worm compost with them in return.

Chapter 8
Maintenance

Now that you've build your worm bin, you have to take care of it.

A worm bin is easy to care for, but it does require some maintenance from you, the worm farmer. In this chapter, you will learn about moisture level, controlling temperature, worm composting with kids, and other critters you will find in your worm bin.

The Correct Moisture Level for a Worm Bin

All worms breathe through their skin. To do this, their skin must stay moist. If a worm's skin dries out, it will die. You may have seen this when an earthworm gets stuck trying to cross a sidewalk in the summer. It ended up dried, shriveled, and dead.

A worm composting bin should never be dry, but there should never be standing water in it either. Ideally, the bedding should be about as moist as a wrung-out damp sponge. The bedding should clearly feel moist. But when you squeeze it in your hand, no water should drip out. Also, when you squeeze it, you should not hear crackling of dry paper or dry leaves.

Too Much Moisture

Too much moisture in your worm bin can make it smelly and muddy-looking. It also reduces the amount of available oxygen for your worms. Too much moisture in your worm bin is usually caused by one of two things: you added too much water when you were first making your bin, or your worm bin got too wet over time by adding high water content food scraps (80%–90% of food is water).

Either way, a worm bin that has too much moisture is easy to fix. Add a couple handfuls of fresh dry bedding. Shredded paper works really well for this, but any type of worm bin bedding that you are using will work. Mix the dry bedding in a bit of water—it will soak up the excess moisture. This will work for any type of worm bin and should be the first thing you try.

If your bin does not have a drain in the bottom, you may need to turn it on its side to drain some of the excess water from the bottom. You'll need to do this a couple times a year if you are using a storage tote-style worm bin. Simply tip the worm bin on its side so that the worm tea can run out one of the air holes that you drilled. Be sure to do this directly either into your vegetable garden or into a bucket so that you can add the nutrient-rich tea to your favorite plants.

Not Enough Moisture

If your vermicomposting system is ever too dry, you need to add some water. The easiest way to add moisture evenly to a worm bin is to use a spray bottle. Simply spray the dry areas of your bin until they are the correct moisture level. You should not have to do this too often.

One way to help regulate the moisture level in your worm bin is to place several sheets of soaked newspaper loosely over the bedding like a blanket. This will help retain the proper moisture level.

The Correct Moisture Level

As you care for your red wiggler composting worms and your vermicompost bin, the moisture level of the worm bedding will fluctuate over time. Use the above techniques to help maintain the correct level of moisture for a worm bin.

Regulating Temperature in a Worm Bin

Red wiggler composting worms tolerate a wide range of temperatures. They can easily be kept indoors or outdoors. This section will provide all the information you need to keep your worms within their ideal temperature range so that they stay happy and productive.

The Ideal Temperature for a Worm Composting Bin

This is easy to remember. Composting worms like the same temperatures that we do. Their ideal range is 60°F to 80°F (15.56 C to 26.67 C). Go to your worm bin. If you are cold, so are they. If you are hot, so are your worms. Even though your composting worms will tolerate temperatures outside this ideal range, do your best to keep the temp within the range because this is when your worms are most productive. (See below for tips on raising and lowering the temperature inside your worm bin.)

When the temperature in your worm composting bin starts to drop below 60°F, your worms will start eating less and mating less. As the temperature continues to drop, they will go into survival mode and huddle together, usually in the center of the bin. If the worm bin temperature drops below 40°F (4.44 C), the worms will start to die off.

If the worm bin temperature rises above 80°F, the worms will start to eat and reproduce less. If it gets above 95°F (35 C), your composting worms will either die or leave in a mass exit—one of every worm composter's worst nightmares! Be careful not to overfeed your worms. If you add too many food scraps to a large worm bin, it will become a hot compost pile and create its own heat, cooking your worms from the inside out.

Since worms prefer the same temperatures that we do, why not keep your worm bin inside? One of the benefits of worm composting is that you can easily keep your worm bin in your kitchen closet, garage, or even under your desk at work, like I do. If your worm bin is too large to keep inside or your spouse is dead set against it, there are ways to worm compost outdoors even in the winter or the heat of summer.

How to Lower the Temperature in Your Worm Composting Bin

Worm composting outdoors in the heat of the summer can be a challenge, especially if you live in a part of the world where it gets really hot. If you live on the equator, please consider keeping your worm bin indoors.

If you live in a milder summer climate, try some of the following techniques to lower the temperature of your worm bin:

- *Keep your worm bin in the shade*—duh.

- *Increase air flow.* You can do this by drilling more holes in your worm bin or simply removing the lid for a time. If you remove the lid, watch out for the family dog digging in the bin or, as at my house, the family chickens helping themselves to a snack.

- ***Make sure the bedding stays moist.*** Composting worms need moist bedding to keep their skin moist so that they can breathe. Heat dries out bedding quickly. Use a watering can or spray bottle to keep the bedding wet. As the water evaporates, it lowers the temperature of the worm bin.

- ***Use a fan.*** Open the lid on your worm bin and point a fan at the bedding. The dramatically increased air flow will increase evaporation and lower the temperature. If you use a fan, be sure to keep the bedding moist because the air moving across the bedding will dry it out faster.

- ***Use a wet towel with a fan.*** For added cooling effect from evaporation, drape a wet towel over the top of your worm bin with the lid removed.

- ***Add some ice.*** When your worm bin is way too hot (like when you have accidentally turned it into a self-heating hot compost pile by adding way too much food scraps), throw some ice on top.

- ***Install a fan.*** It takes a little rigging, but a small computer fan works well. Install it in the lid of your worm bin to aid in air flow. This approach works best in a larger continuous flow-through worm composting bin. You will need an electrician friend to help you with the basic wiring. Want to get really fancy? Add a thermostat that turns the fan on when it gets above a certain temperature.

How to Raise the Temperature in Your Worm Composting Bin

Worm composting outdoors in the winter is possible, but probably not if you live in Antarctica. Before you set up a winter worm compost bin, think about your climate and average temperatures.

Experienced worm farmers willing to build a large outdoor winter worm bed have had great success even in a place as cold as Canada.

Try the following tips for raising the temperature in your worm bin:

- ***Keep the lid on.*** If your winter worm bin has a lid, keep it on to trap heat in. If it doesn't, use a tarp to cover the worm bed.

- ***Provide insulation.*** Insulation can be as simple as surrounding your worm bin with hay bales or as complex as attaching sheets of blue board insulation foam. Any way you choose to do it, providing insulation will help keep your worms warm. I transformed a broken chest freezer into a worm bin. It works great and has built-in insulation! Old coolers work well too.

- ***Build a large winter worm composting bed.*** I have found that larger worm composting systems are less susceptible to atmospheric changes than smaller systems are. Therefore, it is easier to regulate the temperature of a larger worm composting system. This is especially true for keeping a large system warm in cold weather. You can add lots of food scraps at one time to a section without worrying about hurting your worms. Because the system is large, worms can move around and find a spot closer to their ideal temperature.

- ***Use a heater.*** Buy or build yourself a warm worm heater. The heater I use came with the Worm Wigwam and is composed of a thermostat (so that it can turn itself on) and the heating coils used to heat seed trays in a greenhouse. The coil is stapled to a board that lies flat (heat side down) on top of my worm bin. If you are handy, you can make your own. If not, you can achieve the same effect by purchasing seed tray warming mats and placing them on top of your worm bin. Be sure to buy the ones with a thermostat so you don't have to run out in the cold to turn it on!

- ***Put your worm bin next to the outdoor exhaust vent for your heater or dryer.*** My gas pack (heater and AC combined) has an exhaust vent that blows warm air whenever the unit is on (which is often during the winter). Why not use this discharged heat to help keep your worm composting bin warm? This technique will also work with your dryer vent that blows warm moist air outside your house. (Don't try to vent your dryer inside the house, though!)

FREEZE WARNING

Composting worms can freeze! I have had success worm composting outdoors in the winter in North Carolina, but it doesn't get that cold and doesn't stay cold for very long. The above tips will help—but only up to a point. If you get extremely cold temperatures for a long period, you need to move your worms inside to keep them alive.

Leaving Your Worms While on Vacation

Many worm farmers worry about leaving their vermicomposting bin when they go out of town. They do not want a great vacation ruined by returning home to a bin full of dead worms or a house full of worms that have escaped from their worm bin.

So, what should you do to your worm composting bin before you head out of town so your composting worms are safe and happy while you're gone? Should you feed them more? Add worm bin bedding? Hire a wormsitter? It all depends on how long you will be away.

If you are going to be away for less than two weeks, there isn't much extra you need to do. If you plan to be away for longer than two weeks at a time (lucky you!), there are a few simple steps below that will ensure the safety of your composting worms while you are away.

For example, if you are a teacher with a worm composting bin in your classroom, you don't need to take your worm bin home each weekend or even over spring break. But I recommend taking them home over the longer winter break, and certainly over the summer.

Away from Your Composting Worms for Less than 2 Weeks

Here is a simple checklist to follow before you leave town:

- Feed them the day before you leave. Feed them only the same amount that you usually do. Remember, most worm bin problems stem from overfeeding.

- Check the weather forecast. Be sure there won't be any drastic temperature changes that will hurt your worms. If there will be drastic temp changes, plan accordingly.

- Add more bedding if it is low. Your composting worms will eat their bedding. They won't starve.

Away from Your Composting Worms for More than 2 Weeks

Here is a simple checklist to follow to ensure your composting worms enjoy your vacation as much as you did:

- Feed them the normal amount the day before you leave.

- Check the weather and plan accordingly.

- Add more worm bin bedding.

- Get a friend to check on them halfway through your extended vacation. Ask him to feed your normal amount. This works best if you have a friend who is also a worm farmer so that you don't have to train him. But, hey, if you do have to train a rookie, thanks for spreading the worm composting knowledge!

- Consider asking a friend to keep your worm bin at his house while you are away. Again, this is easiest if your friend is a worm composter already. But if not, simply refer him to this book for an overview.

Follow these simple worm composting vacation checklists and your worms will be happy and healthy when you return.

Chapter 9
Other Critters in the Composting Bin

Following are descriptions of good and bad organisms you will find in your worm bin.

In a healthy, productive worm bin, it is perfectly normal and expected to find a few other critters living alongside your composting worms. In my worm bin, I have found rolly pollies (sow bugs), ants, centipedes, millipedes, earwigs, pot worms (very small white worms), slugs, and other critters. These other bugs, insects, and critters won't hurt your worms and are usually a sign of a healthy worm bin because they help with the decomposition process.

A healthy worm composting bin is its own ecosystem. Of course, it will have lots of composting worms, but it will also include lots of other bugs and insects. Most of the other critters you will find in your worm bin are harmless to your worms and to you as the worm farmer. But a few are harmful.

When you find any of these pests, you need to deal with them so that your composting worms don't suffer. Your job as the worm farmer is to know the difference between the good organisms and the harmful ones.

Good Bugs in a Worm Bin

Rolly Pollies (Sow Bugs, Pill Bugs)

Rolly pollies are actually crustaceans and need a very moist environment to live. They love the moisture in a worm bin and eat a vegetarian diet. They will not harm your worms.

Earwigs

Earwigs are often misunderstood insects. They aren't called earwigs because they will crawl into your ear and eat your brain! Earwigs are skinny insects ranging in length from ¼ inch to ½ inch. They have wings but rarely use them. They do have pincers coming out of their abdomen that look pretty menacing. They do not have any venom or poison. If provoked they may use their pincers on you but they aren't strong enough to break the skin. Earwigs are nocturnal, hiding during the day and active at night. Earwigs are proud members of the decomposition club. They help break stuff down and will not harm your composting worms.

Springtails

Springtails are teeny, tiny white insects. You can see them with the naked eye; they look like very small white dots. Springtails get their name from the long (relative to their body size) tail that they keep under their abdomen. When they are threatened (like when you start digging in your worm bin), they push off with their long tail, jumping away. Springtails do not harm composting worms and aid in decomposition.

Millipedes

Millipedes (thousand-legged) are round-bodied and have two pairs of legs per body segment. They are vegetarian and help with the decomposition process. Millipedes are not the same as centipedes (hundred-legged). Centipedes are flat-bodied and have only one pair of legs per body segment. Centipedes kill worms. Millipedes do not.

Pot Worms

Many people confuse pot worms in their worm bin with baby red wiggler worms. Pot worms are another variety of earthworm that lives happily in a worm bin or compost pile. They are very small, white-colored worms. Although pot worms aren't by themselves a bad thing (they also help with decomposition), pot worms thrive in different conditions than red wigglers do. So, if you have a pot worm infestation, your worm bin or compost pile is not set up well for your red wigglers to thrive.

Mites

Mites are common in a worm bin. They are very small and to the naked eye just look like little specs. There are many, many varieties of mites. Most eat rotting organic matter or the poop of another creature. Most mites are a benefit in your worm bin, though an earthworm mite (see below) infestation can be bad for your worms.

Black Soldier Fly Larvae

People often freak out when they first find soldier fly larvae in their worm bin because they look like maggots. They are similar to common house fly maggots in that they are the larval stage of a type of fly. However, they eat any type of food scraps, not only meat. Many people intentionally add black soldier fly larvae to compost their food scraps. I have even set up black soldier fly composters on top of my chicken coop so that the adult larvae crawl out (when they are ready to turn into a fly) and drop right down into the chicken coop. My chickens love eating them and have learned to wait under the composters. A free protein source! Black soldier flies are found naturally in most parts of the world and will most likely find your outdoor worm bin in the summer. They will not hurt your composting worms and can easily coexist in the same bin.

Bad Bugs in a Worm Bin

Centipedes

Centipedes are skinny and flat-bodied with one pair of legs per body segment. *Centipede* means "one hundred-legged." Centipedes are rare in a worm bin and should be removed when found. They are predators that kill and eat earthworms. Centipedes are not the same as millipedes (thousand-legged). Millipedes are round-bodied, have two pairs of legs per body segment, and are vegetarian (they don't hurt worms). Learn to tell the difference and kick those centipedes out.

Earthworm Mites

Most mites that you find in your worm bin are harmless to your worms and help with decomposition. But one species of mite—the earthworm mite—can negatively impact your worm bin. Earthworm mites are reddish-brown. In a serious earthworm mite infestation, your composting worms may refuse to eat because their food is covered in red mites. If this happens, remove and destroy any food scraps visibly covered in red mites. If you want to get rid of even more, set up a trap with some bread soaked in milk. Leave the bread in the bin overnight. Then remove it along with any earthworm mites that have accumulated. Infestations to the point of slowing down or stopping worm feeding are rare.

Ants

Ants have never been a problem in my indoor worm bins. I have had lots of sugar ants in my larger outdoor worm composting systems. An ant infestation is usually a sign that your worm bin bedding is too dry. Add some water with a spray bottle or watering can. If that doesn't work, try sprinkling ground cinnamon wherever the ants are. Ants hate cinnamon. Don't use a poison to get rid of ants because that will hurt your worms.

Roaches

Nobody wants cockroaches in their worm bin, and I have found them to be rare. The presence of roaches in your worm bin depends a lot on the presence of roaches in your house. Roaches can also be a signal that your worm bin bedding is too dry. Try adding some water with a spray bottle. Also, be sure to bury your food scraps well because roaches are attracted to food left out. If you need to, set up a few roach bait traps outside your worm bin. Do not put them inside; they can damage your worms.

Spiders

Most spiders are generally harmless to worms, but they are unpleasant for the human worm farmer. For that reason, I would remove any spiders that you find in your worm bin.

Fruit Flies

If you are a worm composter, you will get fruit flies at some point. They are a pain. Fruit fly eggs often ride into your home on the skin of the lovely fruit you purchase. As the fruit starts to ripen and eventually rot, the eggs hatch, and fruit flies emerge. If not controlled immediately, a small fruit fly problem can easily turn into a large fruit fly problem. Take steps to prevent a fruit fly infestation, and immediately take action if you see fruit flies.

9 Ways to Get Rid of Fruit Flies Fast

1. ***Build a fruit fly trap.*** If you've already been invaded with fruit flies, it's time to trap them and kill them. (More on this later.)

2. ***Get rid of any fruit in your house.*** Remember, fruit flies lay their eggs in fruit skin (they especially like bananas, which ripen quickly). If you are fighting a fruit fly invasion, get rid of all fruit, new and old, sitting on your counter.

3. ***Empty all trash cans*** immediately and clean with bleach. Then leave them out in sun to dry. Fruit flies can live in your trash can as well, especially if you are putting food scraps in your trash can.

4. ***Clean your kitchen counters and sink*** with a diluted bleach spray.

5. ***Toss out your old sponge.*** An old stinky sink sponge with tiny food particles is a great habitat for fruit flies.

6. ***Add another layer of worm bin bedding*** to the top of your worm composting bin. You should never be able to see food scraps.

7. ***Bury all your food scraps well*** when you feed your composting worms.

8. ***Do not overfeed your composting worms.*** If you put too much food in the worm composting bin, it will decompose without the worms eating it. Rotting food scraps make a great habitat for fruit flies. Don't put extra food scraps in your worm bin before you go out of town. Composting worms can eat their worm bin bedding as well. They won't starve if you leave them for three to five days.

9. ***Freeze your food scraps*** before putting them in your worm bin or outdoor compost pile. Keeping your compostable food scraps in the freezer instead of on the counter keeps fruit flies from breeding in them. Freezing food scraps also kills any fruit fly eggs that may have been on your fruit skins.

Make a Homemade Fruit Fly Trap in Under 2 Minutes

My natural fruit fly trap recipe works well, takes less than two minutes to set up, and is made from stuff you already have in your kitchen right now.

Homemade Fruit Fly Trap Materials

- A drinking glass, jar, or cup. I use a clear glass because I like to see how many flies I catch.
- Roll of tape. Any type will do.
- A funnel.
- Some vinegar, wine, or some kind of sweet liquid that attracts the pesky fruit flies.

Homemade Fruit Fly Trap Assembly

- Fill the glass with a half an inch or so of the bait. I used red wine, but vinegar works well too.
- Place the funnel in the glass so that the end is not touching the liquid.
- Tape the lip of the glass to the funnel so that the fruit flies can't escape.
- Leave your awesome homemade fruit fly trap on the counter overnight and wait for the little buggers to find it. Place the trap near the source of your problem: rotting fruit or trash can.
- The next morning, check your fruit fly trap and rejoice.

How This Simple Homemade Fruit Fly Trap Works
Fruit flies smell the sweet liquid bait and fly down into the funnel. Once inside the glass, they can't find their way back out and end up drowning in the bait. I like to use red wine in my homemade trap so at least the fruit flies die happy!

Chapter 10
How to Keep Worms in the Bin

One of the biggest fears of any beginning worm farmer is worms leaving the bin all at once. But you really don't have to worry.

Can you imagine coming home to a mass exodus of 50,000 red worms crawling all over your kitchen floor?! The reality is that there is a *very slim* chance of this ever happening. If you follow good worm composting practices, it will never happen.

Why Composting Worms Might Leave a Bin

We know that composting worms need a dark, moderate-temperature home filled with moist bedding. If they do not have this environment, they will die. Worms know this instinctually.

The only reasons they would leave the bin is because conditions have gotten so bad inside that bin that they have no choice but to bail and look for a better spot. Your job as the worm farmer is to make sure conditions inside the bin stay preferable to conditions outside the bin.

Here are the major reasons your worms would ever stray from their worm bin:

- ***Exploration of brand new worm composting bin:*** When you first add composting worms to a new worm bin, they will crawl all over the inside of the bin checking it out within the first couple days and nights. Usually, they don't leave the bin during this "settling in" stage. But to make sure that they don't, simply leave a light on in the room with the worms. Worms don't like the light, so they will be encouraged to burrow down in the worm bin bedding.

- ***Too much water, not enough air:*** Another reason worms may leave their cozy worm bin is if you allow it to get too wet inside. The worm bedding needs to be moist but not soupy or soggy. Too much water means the excess water fills up all the space in the bedding that is usually filled with air. This lack of air and crawl space for the composting worms could cause them to look for another home outside the worm bin. If your bin is too wet, add some drainage holes or some dry bedding to soak up the extra moisture.

- ***Hot worm bin:*** Composting worms will leave their bin if it gets extremely hot. Be sure your worm bin stays within the ideal temperature range by keeping it either indoors or outdoors in the shade.

- ***Overfeeding:*** Be sure not to overfeed your worms with food scraps. Overfeeding can be a cause of a high temperature. A large worm composting bin can turn into a hot compost pile if you add too much food waste too quickly. Because of the high water content in vegetable food scraps, overfeeding can also cause your bin to be too wet.

How to Ensure Your Composting Worms Never Crawl out of the Bin

Follow these steps and you will never have to worry about a worm escape again:

- ***Provide sufficient bedding.*** Make sure your composting worms have enough moist carbon bedding. Ideally, your worm bin should be ¾ full.

- ***Ensure proper moisture content.*** Too much water and not enough air is one of the few reasons worms will leave their bin. Make sure you maintain the correct balance.

- ***Feed them but not too much.*** Give your worms more food only when they have eaten the last meal you fed them.

- ***Add light.*** Still worried about your worms leaving? Add some lights. We know worms don't like light. If you are really worried about it (which you don't need to be if you've followed the steps above), you can leave a light on all the time in the room where you keep your worms. A strand of Christmas lights decorating the top of the bin works well. For the environment's sake, use LED Christmas lights that don't need a lot of electricity to run.

- ***Split your worm bin every year or so.*** If your worm bin seems overcrowded, consider splitting it in two to start another bin you can keep or give to a friend. I have not found worms to leave a bin because of overcrowding. But splitting the bin will help keep the population in check and will provide you (or a friend) with another worm bin! If you have a healthy, active worm composting bin, you can easily split the worms and bedding in two to make an additional worm bin. Make sure your friend is excited about becoming a worm farmer to ensure the worms aren't neglected. If she needs any help learning about worm composting, be sure to recommend this book!

How to Split Your Worm Composting Bin

- ***Wait until your worm bin is well established.*** If you have a new worm bin, you will need to wait four to six months until the population of your worm bin reaches the limit imposed by the size of the bin. At this time, it is ready to split.

- ***Never split a foul-smelling worm bin.*** A worm bin should never smell bad. If it does, it's a sign that bin conditions are out of balance. Splitting a foul-smelling worm bin only doubles the problem. Address the problem instead by consulting the worm composting troubleshooting guide in Appendix 1 of this book.

- ***Prepare a second bin.*** Either make the bin yourself or purchase a commercially available worm bin.

- ***Move half of the bedding and worms*** from your first bin into the second bin.

- ***Add new moist worm bin bedding*** to each worm bin to bring it up to within several inches of the top of the bin.

- ***Share your second worm bin*** or keep it to double your production. If managed properly, the worms in both bins will grow their populations to the size that the bin allows.

Chapter 11
The Poop:
All About Worm Compost

This chapter is all about the worm compost. You will learn different ways to harvest and use your worm compost so that it has the largest impact on your plants.

One of the primary benefits to worm composting is to produce nutrient-rich worm compost that you can use in your garden. Worm compost (also called vermicompost, vermicast, or worm castings) is simply worm poop. Worm compost is very dark and crumbly and smells like earth or soil. Worm castings alone never stink. Worm compost is alive with microorganisms and nutrients that help all plants grow.

How Long You'll Have to Wait

In my experience, given ideal worm composting conditions, a worm bin started with 1 pound of composting worms (about 1,000 worms) produces usable worm castings in about six months. After that initial harvest, you will be able to harvest a small amount every month or so, depending on the size of your worm compost bin and your worm herd.

However, your specific worm composting situation depends on lots of different factors.

- *Worm bin conditions:* Are you following recommended worm farming practices? Is the bin setup and maintained correctly? If not, the worms will not thrive and be as productive as they could be.

- *Initial worm herd:* How many worms did you start with? Worms will, of course, reproduce quickly. But starting off with more worms will help you achieve a worm compost harvest faster.

- *Size of the worm bin:* Is your worm bin small? Medium? Large? Does your worm herd have room to grow in population?

7 Worm Compost Harvesting Methods

Harvesting your valuable worm compost can be a bit tricky because you need to separate your worm herd from the compost before using it.

Here are different ways to harvest your worm compost:.

1. ***Handfuls:*** The simplest way to access your worm castings is digging down to the bottom of the worm compost bin and grabbing a handful of vermicompost. You can either add it to your garden, worms and all, or pick out the composting worms and then add it to your garden.

2. ***Pyramid method:*** Another option is to wait until most of your worm compost bin has been converted into worm castings (they eat the bedding too) and then dump the whole bin out on a tarp on a sunny day. Form a couple small pyramids of worm castings, and the worms will burrow into the pyramid because they don't like the light. Then carefully brush worm castings off of the outside of the pyramid and set aside. The worms will burrow again. Repeat until almost all worm castings are harvested and you are left with a ball of worms. Refill your bin with fresh worm compost bin bedding, and add your composting worms back to the bin.

3. ***Side-to-side migration method:*** Wait until all the bedding has been processed into vermicompost. Then pull it all to one side of your worm compost bin, leaving the other side open. Fill the open side with fresh worm bin bedding, and begin feeding your worms only on that side. Over the course of a couple weeks, most of your worms will have taken the hint and moved to the new side, so you can pull finished worm compost from the other.

4. ***Screen-and-light method:*** Take advantage of the fact that worms don't like the light and will move away from it. Pull several handfuls of vermicompost out from the bottom of your worm compost bin and spread thinly on a piece of window screen or very fine hardware cloth (1/4 inch or 1/8 inch holes work well (0.64 cm to 0.32 cm)). Then put the window screen or hardware cloth on top of your open bin so that it is resting on the bedding. Shine a light on the vermicomposting bin or leave it outside on a sunny day. The worms in the vermicompost will move away from the light, down through the screen, and back into their worm compost bin. Wait a couple hours for the worms to move. Then simply lift the screen off and use the worm castings.

5. ***Corralling methods:*** You are a worm rancher, so why not set up a corral? Use an old onion or potato bag that has many holes in it (burlap also works well). Fill the bag (the worm corral) with some fresh bedding and some of the worms' favorite food. Tie the bag shut and bury it or place it in your worm compost bin. Be sure the worms don't have any other food in their bin. Wait a couple weeks for the worms to get inside the bag corral. Then simply lift out your worm corral and use your worm castings.

6. ***Cylinder spinning harvester:*** If you have a lot of worm compost to process, consider building or buying a worm compost harvester designed to separate the worms from the worm compost. Check out this video of me using the one that I built.[4] The cylinder is made from ¼-inch (0.64 cm) hardware cloth. Handfuls of worm compost and worms go in at the top. As it spins, the small chunks of worm castings fall through the screen while the large chunks and the worms come out the back. It works decently but is certainly not perfect.

---

[4] http://wormcompostinghq.com/screening-worm-compost-with-homemade-screen

7. ***Flow-through bin:*** My favorite way to harvest worm compost is to use a bin that helps. Build or buy yourself a continuous flow-through worm bin, and harvesting is much easier. Flow-through worm bins usually have some sort of grate holding the bedding up and are open underneath for harvesting. Composting worms move upward as they eat, leaving the worm castings behind. The very bottom should not have any worms in it. Castings fall through the grate on their own because of the worms movement, or you can use a hand cultivator to agitate and knock it through. The Worm Wigwam (reviewed in chapter 5) even has a harvest crank attached to a blade that sits on top of the grate and scrapes off the bottom layer of worm compost.

5 Ways to Use Vermicompost in Your Garden or Lawn

Worm compost is an excellent organic fertilizer or soil amendment that will make any plant healthier. This section will teach you five easy ways to use worm compost in and around your home to make the most of its benefits.

1. ***Top-dress plants with worm compost.*** The simplest way to use vermicompost is to add it to the top of the soil around your plant. This method for using worm compost is called top-dressing. Top-dressing benefits indoor as well as outdoor plants. Simply use your hands or a small shovel to put the worm compost on the soil directly around the stem of the plant. This is a great way to use worm castings in your garden because of their potency and because you may not have enough worm castings to dig it into the soil in large quantities.

2. ***Add worm compost to your garden when you turn the soil.*** Between growing seasons, gardeners often turn over their soil to prepare for the next growing season. This is a great time to add a large quantity of vermicompost to your garden. Simply cover your garden space with 2 or 3 inches (5 to 7.6 cm) of worm compost. Then fold it over as you normally would, using a tiller or shovel. A drawback of using this method is it takes a large volume of worm compost. Your worm composting bin may not produce enough vermicompost to add it to your garden in large quantities. If that is the case, you can either purchase some bagged worm compost online or make a larger worm bin.

3. ***Use worm compost in a plant starting mix.*** Whether you are transplanting a potted plant or starting vegetable plants indoors, the plants will benefit if you add some worm compost to your growing medium. You can purchase a potting soil or seed starter mix that you can add your vermicompost to, or you can make your own growing medium. (In the next section, I will share my personal recipe for seed starting mix with worm compost.) Because you want your plants to get the best start possible, this is a great use of your precious nutrient-rich worm compost.

4. ***Make a vermicompost tea.*** Compost tea is a tea made from steeping compost instead of tea leaves. It isn't appetizing for you, but your plants will love it! The nutrients in the worm compost diffuse into the water turning it into liquid gold. (I will cover making and using worm compost tea a little later in this book.)

5. ***Top-dress your lawn with worm compost.*** Just as your houseplants and vegetable garden benefit from the use of worm compost, so will your lawn. To apply the worm compost, simply spread it thinly over your lawn so that it doesn't cover up the grass but, rather, falls down between the blades. You can broadcast the vermicompost over the lawn by hand or use a push seed or compost spreader. This is the same device that you use to spread seed over your lawn. If you plan to use the spreader, be sure that the worm compost is fine and dry; otherwise, it will clog up the spreader. Because this method requires a large amount of worm compost (which you may not have), a more effective use of your precious homemade worm compost may be to apply the worm compost to your lawn in trouble spots only. If you have a couple bald spots in your lawn, give them a boost with the addition of worm compost.

Using Worm Compost in Seed Starter Mix

Worm compost is an extremely nutrient-rich organic fertilizer. You should be using it in your seed starter mix to get your baby plants started off right. Here is my recipe for homemade seed starting mix with worm compost.

My Recipe for Homemade Seed Germination Soil with Worm Compost

- ***⅓ worm compost:*** The most important ingredient! You can purchase worm castings online or in a store. But—as I've done throughout this book—I encourage you to try making your own. Because you will want your plants to get the best start possible, this is a great use of your precious high-powered worm castings. Dried and screened worm compost is much easier to work with.

- ***⅓ coir:*** Coir is a renewable resource made from coconut husks that provides the same benefits as peat moss. Coir is light and fluffy, soaks up water well, and provides room for roots to grow. Coir also makes a great worm bin bedding material.

- ***⅓ vermiculite:*** Vermiculite is a mineral that has been super-heated until it expands or pops. It is ideal for seed starting mixes because of its aeration and water-retention capacity. Adding vermiculite to your seed starting mix will make it more fluffy and help it hold in moisture. Moisture is extremely important because once a seed gets wet, it is turned on. If it dries out, it will die.

I use a large plastic cement mixing tub to blend all three of the ingredients together in whatever volume I need. Mix thoroughly until the worm compost seed starter mix is a uniform color.

Too lazy to make your own seed starter mix but still want the benefits of worm compost? No problem. Simply purchase a bag of seed starter mix from the store and add your own worm compost.

A FARMER'S TIP

After watching me splash my seeds out of their pots the first time I watered them, a farmer friend showed me that he soaks his seed starting mix with water before adding it to the pots. Presoaking the mix makes it easier to add to the seed trays and keeps you from splashing the seeds out when you water them—duh.

Worm Compost Tea

Compost tea is a tea made from steeping (or brewing) compost instead of tea leaves. It isn't appetizing for you, but your plants will love it! The nutrients and beneficial microbes in the compost diffuse into the water, making liquid fertilizer for your plants. Some gardeners refer to compost tea as liquid gold because it is such a powerful organic fertilizer for any plant. There are two ways to make compost tea: the basic method (simple steeping) and the aerated brewed version.

Worm Compost Tea Versus Leachate

Worm compost tea and leachate are different. Leachate is the liquid that leaks out of the bottom of your worm bin. Compost tea must go through an oxygenated brewing process.

Leachate is the water released from the cells of decomposing food scraps. The digestive tract of a composting worm is really good at removing harmful pathogens. Unfortunately, leachate is usually not consumed by the worms and, therefore, can potentially contain toxins harmful to people. I recommend never using foul-smelling leachate; this can be a sign of contamination.

I use leachate in two ways: add it directly to non edible plants and pour it on my hot compost pile that will heat up to kill any harmful pathogens.

Basic Worm Compost Tea Recipe

To make a batch of basic worm compost tea, all you really need to do is soak some vermicompost overnight in water. I prefer to use a makeshift tea bag because it makes it less messy, but it's not necessary.

Here is the basic worm compost tea recipe that I use:

- Find something to use such as a tea bag, old T-shirt, pantyhose, or cheesecloth.
- Fill your homemade tea bag with worm compost and tie it off at the open end.
- Submerge the worm compost tea bag in a bucket of water. I use a 5-gallon (18.9 L) bucket, but any size bucket will work.
- Let it sit overnight. In the morning, the water should be light brown.

- Because the beneficial microbes in the worm compost will start to die off, water your garden first thing in the morning for best results.

- Remove the worm compost tea bag from the bucket, cut it open, and add the worm compost to your garden, your worm compost bin, or your hot compost pile.

Aerated, Brewed Worm Compost Tea Recipe

To brew a batch of aerated worm compost tea, follow roughly the same procedure as the basic worm tea recipe, except you will be introducing a sugar source and an aeration device. The sugar and aeration wakes up, feeds, and increases the population of the beneficial microorganisms living in the worm compost, making this method the absolute best for your plants.

Here is my aerated, brewed worm compost tea recipe:

- Put roughly 1 gallon of finished worm castings (without a tea bag) into a 5-gallon (18.9 L) bucket. I never measure; just use a shovel full of vermicompost.

- Add 4 gallons of water. Rain or well water is best because it is not chlorinated.

- Add 1 ounce of unsulfured molasses to provide a food source for the beneficial microorganisms living in the worm poop. You can use almost any sugar source.

- Stick the bubbler (airstone) end of an aquarium aerator down to the bottom of the bucket and turn it on. Let the brew sit for three days, stirring occasionally.

- You may want to strain the worm compost tea before using.

- For best results, use the brewed worm tea immediately.

- Tip for even better results: follow the worm compost tea recipe above, but add 2 cups (473.18 ml) of alfalfa pellets (rabbit food), for some extra nitrogen in the worm tea brew.

How to Use Worm Compost Tea

Now that you have made a batch of worm compost tea, try one of the methods below in your garden or home. If you have a large garden or many houseplants, dilute worm compost tea with water to cover more space.

Check out these uses for worm compost tea:

- Water your garden as you normally would.

- Water your houseplants.

- Water seedlings or baby plants.

- Cover the whole plant with worm compost tea, including the leaves. Many people believe that the beneficial microbes in worm tea help protect the plant from diseases.

- Serve delicious worm compost tea at your next tea party.

- NOTE: one of these is a joke.

Chapter 12
Worm Composting for Kids

Whether at home or at school, teaching kids the value of worm composting is fun for both the kids and for you.

Kids love worms! Teaching worm composting to kids is also a great way to teach life cycles, biology, conservation, and many other sciences. Worm composting with your kids can also dovetail nicely with gardening with your kids.

Worm Composting at Home

- Always follow good worm composting principles.
- Consider using a smaller plastic bin or tub. A shoebox-size tub will be easier for kids to handle and move around.
- Use a clear plastic container so that kids can see into their worm compost bin.
- Have a parent worm farm as well. It is a good idea to also have a larger worm compost bin in your house in case something happens to your kid's worm farm. You can easily replenish the worm population if you need to.

Worm Composting in the Classroom

Because red wiggler composting worms do very well inside, a worm composting bin is a great educational addition to a classroom. Teachers can use it as an earth sciences teaching tool while composting veggie food scraps from snack time.

The worms will be fine over the weekend, no need to come in and check on them. Just be sure to take them home over the summer!

Worm Composting Activities for Kids

- Identify an adult worm, a baby worm, and an egg.
- Do a "what do red worms like to eat more?" experiment. Place two foods in the bin and see what they eat first.
- Conduct a worm head count. Pull out ¼ of the bedding in your vermiculture bin and count each worm that you find. Then multiply by 4 to estimate how many worms you have total. Do this a couple times a year to see how the population changes.

- Keep worm stats. Measure the length and weight of one worm.

- Time them to see how fast they eat. Add some type of veggie food waste. Then see how long it takes your worms to consume it. Do this experiment again after your worm population changes.

- Use the worm compost (worm poop) to grow some vegetables in a garden. Use vermicompost on some plants and none on others to see the difference it makes.

A Final Word

Earthworms are, by nature, composters. Their role in the ecosystem is to help convert unwanted waste (rotting organic matter) into a usable form (worm castings). Worms have been doing this work for thousands and thousands of years. Your role as the worm farmer is simple: keep your composting worms happy so they do their job well.

I believe that there is no such thing as waste. Everything that we produce that is viewed as "waste" can be recycled, repurposed, or re-used in some way. This belief has led me to try worm composting for the first time so I could stop throwing food scraps in the trash can.

I believe the natural world is balanced. The concept of waste does not exist in nature. Everything is connected. Waste from one system is the input/raw material for another organism or system. We need to let worms do what they do best.

You now have the knowledge. Now, go start your worm bin. If you get stuck and need help, reach out to me at www.wormcompostinghq.com.

Thanks for reading my book. If you enjoyed the book, please spread the word by

- Adding a short review of the book on Amazon. Scroll down and click on the "write a customer review" button.

- Clicking the "like" button on the Amazon page.

- Tell your friends in person or by sharing this book on your social media platforms.

Many thanks!
Henry

Appendix 1
Worm Bin Troubleshooting

Problem	Possible Cause	Things to Try
Worm bin smells bad	• Overfeeding • Uncovered food scraps • Too much liquid/no drainage	• Feed less. • Cover all food scraps with several inches of bedding. • Drain off excess liquid. • Mix in dry bedding.
Worm bin attracts fruit flies	• Exposed food scraps • Overfeeding	• Bury all food scraps. • Build a fruit fly trap.
Worm bin is too wet	• Overfeeding • Worm bin started too wet • Not enough air flow • No drainage	• Feed less. • Mix in dry worm bin bedding. • Add ventilation holes. • Add drains.
Worm bin is too dry	• Dry bedding added • Too much air flow/ventilation	• Spray worm bin with spray bottle of water. • Add several sheets of soaked newspaper to the top of your bin to act as a moisture blanket.
Worms are leaving the bin	• Extremely bad conditions inside bin (This is rare.)	• Correct bin conditions. Worms will leave only if things get really bad. See other solutions. • Shine light on bin to encourage worms to burrow into bedding.

Problem	Possible Cause	Things to Try
Composting worms are dying	• Extremely bad conditions inside bin (This is rare.) • Way too wet or way too dry • No food • Extreme temperatures: way too hot or way too cold	• Correct bin conditions. Worms will leave only if things get really bad. See other solutions. • Correct worm bin temperature.
Rodents in your worm bin	• Outdoor worm bin with exposed food scraps/overfeeding	• Move worm bin inside. • Keep worm bin sealed. • Reduce feeding amount. • Cover all food scraps with bedding.

Appendix 2
Worm Composting FAQ

After years of home worm composting, as well as medium industrial scale (about 12 pounds (5 kg 443.11 g) per day) worm composting, I have compiled a list of frequently asked questions. These questions were asked by adult, kids, teenagers, people interested in worm composting, people who had already started worm composting, and people completely grossed out by the thought of vermicomposting.

Why do you have worms?

I worm compost for two reasons: to responsibly dispose of my organic food scrap waste and to make nutrient-rich worm castings that help me grow more food in the garden.

How long do worms live?

Studies have shown that composting worms can live four to ten years kept in a worm bin or a laboratory. Out in nature, worms typically live one to two years because of predators, temperature changes, and other dangers.

Do worms have teeth?

Composting worms do not have teeth. Instead, they grind their food in very small gizzards. Without teeth, worms cannot take a bite out of food. They need to wait until the food begins to rot or break down so that it is soft and wet enough for them to suck off with their very small mouths.

What is a gizzard?

The gizzard (birds have them too) is a small sack early in the digestive tract (which for a worm runs the entire length of its body) that contains very small bits of grit or sand. The food passes through the gizzard and gets ground up by the grit.

Are there boy and girl worms?

Nope. Worms are hermaphrodites. All worms have both male and female reproductive organs.

Can worms mate with themselves?

No. A single worm cannot reproduce by itself. Even though worms have male and female reproductive organs, they need another partner to reproduce.

How do worms make babies?

The reproductive act, which lasts for around three hours, begins with two mature worms giving each other a hug. They line up their clitellum (the thick band near the head of the worm) and then hold on to each other's bodies with tiny hairs called setae. During this hug, the worms swap reproductive seminal fluids.

Next, the worms secrete mucus rings around both their bodies. The visual effect is that it makes the worms look like they are tied up with very fine fishing string. These mucus rings are the beginnings of the shell of the worm cocoon.

When the worms separate, the mucus ring slides off each worm, collecting fertilized reproductive seminal fluids as it moves along the worms' bodies. When the mucus ring gets to the end of the worm, the ends of the mucus ring seal themselves, creating the cocoon that contains all the necessary reproductive material. The cocoon then separates from the worms to develop.

How quickly do worms reproduce?

It takes a baby worm only nine weeks to reach maturity and start reproducing. Worms reproduce by creating small tan-colored cocoons. Each cocoon holds two or three tiny baby red wigglers. So, given ideal conditions, enough food, and enough space in the worm bin, you can expect your worm herd to double every three or four months.

How much can red wiggler worms eat?

Red wigglers are voracious eaters. Depending on the conditions in the worm bin, they can eat between ¼ and ½ of their weight every day. So, if you have 1 pound (453.6 g) of worms (roughly 1,000 worms), you can expect them to eat ¼ pound to ½ pound each day (113.4 to 226.8 g) —under ideal conditions. To determine how many worms you need to eat all the food scraps your family produces, do a food waste audit.

How long does it take composting worms to make usable castings?

It takes around six months before you are able to harvest vermicompost for the first time. After that, you will be able to harvest a small amount every month or so, depending on the size of your worm compost bin and your worm herd. Want vermicompost faster? Start with more composting worms!

Which worm bin do you recommend for beginners?

If you want to build your own bin, I recommend making a simple tub-style worm bin. Simply drill some air holes in the sides of the tub, fill it with moist worm bin bedding, and add composting worms.

If you would rather purchase a worm bin, I recommend the Can-O-Worms. It is my favorite tray-harvesting-system worm bin.

How do you harvest the castings?

Most of my worm bins are flow-through worm bins where the worms are fed on top and castings (worm poop) are harvested from the bottom. I also use light and a window screen to separate worms from the compost. Worms naturally move away from light, so spread your worm compost in a thin layer on a window screen placed on top of your open worm bin, and shine a bright light on your worms. They will move away from the light down through the screen and back into your bin.

How do you use the castings?

Worm castings make great nutrient-rich natural fertilizers that will make any plant grow. You can use them in many ways. Sprinkle them around the stems of existing plants; make a worm casting compost tea out of them to water plants; add them to a seed starting mix—just to name a few.

Where do you get the composting worms?

Believe it or not, you can purchase composting worms online and have them mailed to you. Here is a worm seller that I trust.[5]

How many worms should I start with?

I recommend starting a homemade bin of this size with 1 pound of composting red worms. If you are ready to get started, see below for a link to a worm seller I trust. The worms will multiply, so the amount of red worms you need to start with really depends on how productive you want your homemade worm bin to be right away, and how much you are willing to spend. If you would like to know exactly how many worms are needed to compost all the food scraps that your family produces, then follow my instructions for conducting a family food waste audit.

Does a worm bin stink?

Rotting food scraps? Worm poop? Doesn't it stink? As a worm farmer, you will be asked these questions often. The answer is: worm bins should never stink. A bad smell is an indicator that something has gone wrong and needs to be fixed.

---

[5] http://wormcompostinghq.com/buy-composting-worms

Does a worm composting bin attract other bugs?

In my worm bin I have found rolly pollies (sow bugs), ants, centipedes, millipedes, earwigs, pot worms (very small white worms), slugs, and other critters. These other bugs won't hurt your worms and are usually a sign of a healthy worm bin. If your outdoor worm bin is not maintained properly, it can attract pests you don't want, such as ants and flies. Proper maintenance or moving the worm bin inside usually corrects the pest issue.

Does a worm bin attract rats?

Nobody wants rats hanging around their house. A well-maintained worm bin will never attract rats.

Can I keep composting worms outside?

Sure. You can keep a worm composting bin outside. Just be sure to have a plan for controlling temperature so your worms don't freeze or overheat.

Do worms have eyes? Can a worm see?

Worms do not have eyes. They don't see in the way that we do. They are extremely sensitive to light and move away from any light source they encounter. Red light bothers worms less, so if you want to observe them in a more natural way, use a red light rather than a white light.

How do worms breathe?

Worms breathe through their skin, and their skin must be moist to allow them to breathe.

Why do they have to be in moist bedding?

If a worm's skin dries out, it will die. You may have seen this when an earthworm gets stuck trying to cross a sidewalk in the summer. It ended up dried, shriveled, and dead.

Does a worm have a front and back?

Yes. The mouth end is the end near the clitellum, the thick band that is present on sexually mature worms.

Do worms die in a worm bin?

Yes. Worms complete their lifecycle and die in a worm bin. However, worm farmers should rarely, if ever, see dead worms. The other critters in the worm bin do a great job of decomposing any dead worms.

Can worms regenerate?

Yes. But their power is limited. They can regrow sections of their body after injury.

If you cut a worm in half, will both parts grow back?

Nope. There is a chance that the side with the head (and five hearts) will regrow a tail, but most likely the worm will die.

What should I use as worm bin bedding?

You can use almost any carbon source as worm bin bedding, but some worm bedding material works better than others.

What is the correct moisture level for a worm bin?

Worm composting bins should never be dry and should never have standing water in them. Ideally, the worm bedding should be at about 80% moisture. The bedding should definitely feel moist, but when you squeeze it, no water should drip out. Also when you squeeze it, you should not hear crackling of dry paper or dry leaves. When your worm bin bedding is at the correct moisture level, it should remind you of laundry right when you take it from the washing machine. The clothes are thoroughly saturated with water but not dripping at all.

What temperature do worms prefer?

They can tolerate a fairly wide range of temperatures. Red wigglers are most efficient (eating, pooping, making babies) at temperatures that we humans prefer, roughly 60°F to 80°F (15.56 to 26.67 C).

About the Author

Henry Owen fell in love with nature as a child at summer camp and on family camping trips. Henry believes that all people need to be connected to nature and that all children deserve opportunities for imaginative, exploratory, nature-based play. After many summers as a camp counselor and then assistant director, Henry earned a master's degree in education and taught second grade. In 2009, he co-founded Friendship Gardens, a nonprofit project that grows food for Meals On Wheels in Charlotte, North Carolina. Now, he spends his days as executive director of the Nature Discovery Center[6] in Bellaire, Texas. He is a husband, father, educator, adventure play advocate, garden nerd, and worm composter. Follow Henry on Twitter @rhenryowen.

[6] http://www.naturediscoverycenter.org

Don't Forget your Free Gift

If you are brand new to worm composting it is tough to know what equipment you really need. *Inside my Worm Composting Toolbox* is my personal list of essential tools and supplies to make worm composting easy.

You can download this free report at
http://www.wormcompostinghq.com/toolbox

Grab it now!

Once again, thanks for your purchase and happy worm composting